GW00725923

AGRICUTURAL HAND TOOLS
1580-1660

Rob Stuart

Stuart Press
117 Farleigh Road Backwell, Bristol
1st Edition 1995
Copyright: Historical Management Associates Ltd 1995
ISBN 1 85804 085 X

Hand Tools

Introduction

The objective of this volume is to identify the tools used, their nature as in dimensions, materials, method of construction, storage location and use.

This 1st series publication draws primarily on written material particularly farming books of the period by authors such as Gervase Markham, Thomas Tusser and Henry Best, a selection of published probate inventories, particularly from the Bristol area, Devon and Banbury, Randall Holmes Academy of Armory and a variety of woodcut illustrations.

Probates have to be used with care. The absence of a particular item often means that it has been lumped under a group category, for example the commonly found term husbandry tools, this is particularly so with relatively inexpensive items. Individual probate inventories vary greatly in the detail they give and prices for second hand objects seem exceedingly variable, probably a function of the amateur assessors estimate or state of dilapidation rather than representing the true new cost of the object.

This volume covers hand tools. It excludes fixtures and fittings and also equine equipment such as combs and scissors which will be covered in latter volumes.

Contents

Digging	4
Pick and pickaxe	4
Mattock	5
Stubbing Hoe	5
Grubbing Axe	5
Spade	6
Hack	7
Paring Spade	8
Trenching Spades	9
Shovel	10
Dibble	11
Iron Bar	12
Weeding	13
Cutting	14
Scythe	14
Sickle	14
Hooks	15
Hatchet and Axe	16
Bills and Cleavers	17
Saws	18
Sheers	19

Splitting
    Hay Hook and Spade 20
    Wedges and Beetles 20
Sharpening
    Strickle 21
    Whetstone 21
    Grindstone 21
Transporting
    Wheelbarrow 22
    Hand barrow 22
    Well Equipment 22
Gathering, Spreading and Flattening
    Rakes 23
    Mauls 24
    Forks 25
Beating
    Flail 27
Sorting
    Sieves 28
    Winnowing Sheet etc 29
    Measure 30
    Bags 31
Climbing
    Ladders 31
Carpentry Tools 32
Miscellaneous 33
Holmes and Tussers tool lists 35
Bibliography and Abbreviations 38

Digging

A number of digging tools are mentioned in period texts. These include Shovels, Spades, Pickaxe and Mattock. Iron bars were probably also used in digging were .

Pickaxes

Pickaxe, Pick or Paviers Pick [RHAA3/8/342]
One was owned by a Leicestershire labourer [LP29] others by a shearman [IAP13] and a husbandman [IAP19]. One was kept in the close by a Gloucestershire Husbandman [CP1]. Tusser indicated that the pickaxe might be found in the cartshed [TT32].

The paviers pick or pick axe had "a long head and back part that it may strike deep into the ground".

Pavier's Pick or Pick Axe
[RHAA3/342/70]

Ancient [in 1688] paviers picks came in various forms:

Type A: the handle could "go into a socket at the head end of iron"
Type B: it might have "a short flat face"
Type C: A Paviers pick, similar to a military pick the difference being that the shaft went through an eye or socket in the iron and fixed there.[RHAA3/8/342]

Type A 72,
Type B 73,
Type C 74.
[RHAA3/342]

Pick or Pioneer Pick [RHAA3/8/337] Military Pick.
The head of the military pick went through the handle and was wedged there [RHAA3/8/342]. The following pickaxes were ordered for the new modle army in 1645:

[RHAA337/45]

500 of Tho Hodgekin att 2s 8d p peece (32d each)[Mung80]
500 pickaxes att 2s 9d p pce (33d each)[Mung71]
400 - Hodgskins of the Tower at IIs Xd p Piece (34d each) stronger sort wth ash helves
    [Mung113] Ash helves were ash handles [Web].

4

Pick

The term pick alone could be applied to an item with a head sharp at one end and with an eye for a handle at the oposite end [RHAA3/8/344]

Edward Roods whose profession was "Grubber" had "two picks for a grubber to woorke with" [Hereford and Worcester Record Office Probate Inventory Dudley 1636]. 3 were kept in the close by a Gloucestershire Husbandman [CP1]

Pick
[RHAA3/344/85]

Mattock

Mattocks were recorded in the possesion of labourers [BP236] Husbandmen [BP235,262] and a glazier [BP270] among others. They were common in Devon, where, given the tendancy for incomplete details, it was probably a rare working farm that did not have at least one and one had 4 mentioned [DP83]. One was found in the back room of an Essex Yeoman's house in 1638 [EP14]. One was located in the hall of Thomas Turnor husbandman of Cakebole, Worcestershire in 1640.

This is an instrument by which labourers do dig and sink into the Earth withal when it is hard and stony, or a clay like substance; the one end of it being broad, and the other sharp pointed [RHAA3/8/342]

Holmes illustration of a
mattock is unclear
[RHAA3/342/71]

These Items are probably forms of either mattock or pick axe:

Stubbing Hoe

"a toole where with labourers stub roots out of decai'd woodland grounds" shaped like a carpenters adze, possibly the same as a grubbing axe. [GMHM183]

Grubbing Axe

One was found along with a mattock in the back room of an Essex Yeoman's house in 1638 [EP14].Grubbing axes and stubbing hoes may be simmilar items.

5

Spades

A number of varieties are mentioned. All the different varieties of spades and shovels in Mungeam were ordered from turners implying that the handles were turned. The heads of the handles could be either t shaped or triangular as per the illustrations although Holmes indicates all spades had T handles: "There is another kinde of Shovel used for Gardening which hath the Handle crosseways like a Spade and is shod square with Iron." [RHAA3/8/331] He also implies that the key difference between a spade and a shovel was that the shovel had a square end

Spades are specified in labourers and husbandmens inventories from Banbury [BP229,236,256,262]. A widow owned a spade valued at 8d [BP283]. A spade was located in the Soller over the parlour and another in the chamber in the inventory of Thomas Turnor husbandman of Cakebole, Worcestershire in 1640, while a Leicestershire husbandman kept a spade and a shovel in the kitchen [LP27] and two were kept in the chamber over the kitchen by a Gloucestershire Husbandman [CP1].

The parts of the spade were: The shank or Stalle

The Sole or broad part on which the Iron is fixed.

The Shoe, or Spade Iron. [RHAA3/331]

Spade Iron
[RHAA3/331/2]

Ordinary Spades
Parliament "agreed with John Wallington and John Coate Turnor for Ordinary spades 900 att 15d each" [Mung 70] for the new modle army. This is the same price as the 100 Steele Spades ordered from " John Grace Turnor att 1s 3d p peece (15d each)" [Mung79] and these may be similar tools. The Ordinary spade was probably what Markham ment by digging Spades [GMHM203]

Variations on the ordinary spade were foreign designs ordered for the New Modle:

Dutch Spades
Agreed with John Wallington and John Coate Turnor for Dutch spades 100 att 3s p pce (36d each) [Mung70]

Flemish Spades
500 Flemish spades ordered by the new Modle Army 1646 [SP28/140/117]

Evelyn

6

Hack

Holmes gives turf spade as an alternative name for this tool although there appears to be a different item generally known by that name.

The hack was shaped like an adze "it is to cut and flea up the surface of anything flat, or of the earth into Sods, Turfs or fleaces, to lay on the ridges of thatched houses in country towns or small cotageş, when they are first cut out by a shovel. This instrument is like a carpenters adze, for as that chips the wood, so this flees up the Earth both with Moss and Grass, and Morish ground." [RHAA3/9/390]

2 hacks along with a mattock costing a total of 2/8d were listed in the possesions at Tatton hall [TPP] Markham talks of hacks of iron, well steeled and reasonable sharp [GMFH5] while Holmes may be reffering to hacks when he talks of turf spades broad and sharp edged, set into a wooden handle like a halbert [RHAA3/7/327,328]

Hacks seem to have 2 uses, firstly after an area of turf or moss had been cut into strips as it lay on the ground the hack could be used to roll it up.[RHAA3/9/390] Alternatively after ploughing clay ground they could be used to chop up any surviving turf in the ridges or at the margins of the field. 1 good hacker [labourer] could cut more than half an acre a day [GMFH5]

Markham talks of a similar tool shaped like a carpenters adze, except twice as big and broader and thinner than a stubbing Hoe, "a toole where with labourers stub roots out of decai'd woodland grounds" with a shaft at least 4 feet long used for striping the green turf between the hills in a hop garden [GMHM183]

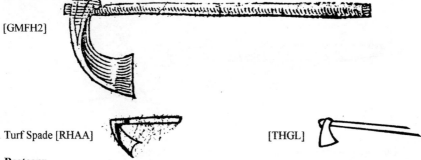

[GMFH2]

Turf Spade [RHAA]   [THGL]

Beateaux

Like a large adze used for breaking up sods for burning [DP46]

Marling Nogger

A Marling Nogger worth 12d was listed in the possesions at Tatton Hall Cheshire [TPP]. A Nauggor was found in the Soller over the parlour in the probate inventory of Thomas Turnor husbandman of Cakebole, Worcestershire 1640.

Paring Spade, Dividing Iron, Turf Spade [RHAA3/9/392]

Markham talks of a paring spade as an alternative to the adze like tool above for use in hop gardens or gardens for stripping turf and roots[GMHM183,203]. A Gloucestershire husbandman owned a paring iron in the chamber over the kitchen [CP1]. This may be another alternative name for the paring spade or dividing iron. Markham's diagram appears to show a typical composite spade with a wooden blade and an iron tip and sides. The bit or shoe was twice as wide as ordinary spades and an inch broader at the point than at the upper end [RHAA3/9/392]

A deepe turffe spade worth 8d was listed in the possesions at Tatton Hall [TPP]

Pareing Spade or Hoe [GMHM183]

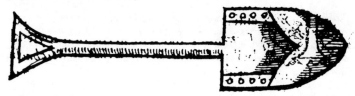

Paring Shovel [GMFH27]

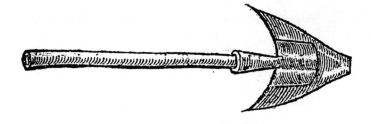

Turf Spade, Paring Iron
or Dividing Iron
[RHAA391/137]

Turf Spade
[RHAA391/140]

8

Trenching Spade
Holmes Trenching Spade "by which grounds are drained...the bottom or shoe of a trenching spade. It is all Iron and put on the staff or staile with a good strong socket. The spade shoe must be made with two sides or Langets up from the end of the bit, as if they were two strong Knife Blades set to turn or look upwards with their points upon a common spade, from the end of the spade bit. If it be made exact with the force of a man pushing it foreward, it will work forth the coar and furrow clearly, and make a trench at one tyme. [RHAA3/9/393]

Fen Labourer using a trenching spade.

Trenching spade shoe .
[RHAA391/140]

Trenching Gouge
Made after the manner of a spade but that it riseth up in the sides, and is round in the Sole or Shoe. The use of it is the very same to the use of the turfing spade, to make trenches in Moorish wet lands to draw away the waters. [RHAA3/9/393]. This seems to have had a curved cross section to create a rounded groove rather than a straight cut.

Trenching Goudge
[RHAA391/139]

A goudge was listed among the possesions of a Gloucestershire Husbandman in the chamber over the kitchen [CP1] and may have been a trenching goudge.

The next three tools are similar and may be identical to some of the above.

Skuppat
Used by marsh men for amking narrow ditches for drainage work [TT33]. A Norfolk blacksmith had 10 skuppet shafts valued at 4.5d each [WP28]

Skavel
Used by marsh men it was a spade with the sides slightly turned up, used for draining and cleaning narrow ditches [TT33,332].

Didall
Triangular spadeor iron scoop for clearing ditches [TT324]

Shovels

"Shovells att 15d each" [Mung70] cost the same as ordinary and steel spades and were proably made of similar materials. Shovels were recorded in the possesion of labourers [BP236], Husbandmen [BP256x2,CP25x2], Yeoman/Husbandmen [BP278] and a Glazier [BP270] in Banbury. A Leicestershire husbandman kept a spade and a shovel in the kitchen [LP27] and a Devon Yeoman 2 shovels in the stable or its hay loft [DP55]. They were quite commonly mentioned in Devon where one farm had 3 of them [DP83]. They seem far more common in Devon than spades which may be a semantic variation or indicate husbandry differences.

Garden Shovels
There is another kinde of Shovel used for Gardening which hath the Handle crosseways like a Spade and is shod square with Iron. [RHAA3/8/331] A Devonshire gentleman owned among other common tools "1 iron shovell" [DP59] another Devon farm had one Ironde shouelle [DP23].

Round Foot Shovel
There is another kind of Shovel that is square at the bottom, rounded off at the shank with an head like a spade [RHAA3/8/331]

Shovel with a spade handle called by some a Flat or Staight Soled Spade [RHAA3/9/392]

Spit Shovell
Spit shovels occur in the Bristol area in the possesion of tailors [IAP4x2], Husbandmen [IAP10,19x3] a maiden [IAP23] and a miller [IAP38].One was valued at 1/- [IAP10].

Casting Shovel
In Barn [TT31]   Shovel End [RHAA3/337/95]

Corn Shovel or Malt Shovel
The Handles of these kinde of Shovels used about Corne have them ever three square, or cornered, and are without Iron shooing. [RHAA3/8/331]: Possibly similar to casting shovel.

Corn or malt shovel
[RHAA/3]

Crooked Shovel
Crooked shovels occur in the Bristol area in the possesion Husbandmen [IAP19]

Straight Soled Shovel
[RHAA3/391/136]

Dibble
Dibble or round stick for planting seeds [GMHM16-17]

Dibblers in use

GMFH THGL

Spades or Shovels in Use

GMFH THGL

Iron Bars

When iron bars are mentioned in probate inventories and farming texts little indication of their use is given. The following military text indicates that they were probably akin to modern wrecking bars possibly 5 or 6 feet long and used to lever out rocks while digging:

"In 1626 it was directed that for every 100 trayners (Trained Bands Men?) there should be 10 pioneers from the Hundred to be provided with 12 pike axes, 12 spades, 12 shovels, 6 iron bars, 6 axes, 6 hatchets, 2 tent saws, 4 hand saws, 12 small baskets to carry earth, 12 bills to cut wood, 10 berriers (augers) of several sizes [Cot16]."

A large wrecking bar rather than a smaller hooked crow bar is far more practicle for the tasks being performed by pioneers as implied by their other equipment and the role they preformed of clearing a way for the armies through hedgerows and difficult terrain [WIW827].

Iron bars were recorded among the possesons of husbandmen in Banbury [BP235,262] and were common in Devon probate inventories. They are normally found among tools like shovels but occasionaly with kitchen equipment when another form and use may be implied. One was valued at 4/- in Devon [DP111] one was kept in the milk house [DP140] and another kept in the close by a Gloucestershire Husbandman [CP1].

Weeding

Gardners Weeding Dug
"made with a taper fork, and a cross bar of iron, some 6 or 8 inches above and then hath a strong socket into which is fixed a shaft with a spade head as thick or thicker than the spade shank. The cross is for the foot of the workman to force it into the earth, on each side of the weed root and so having hold of it draws it out of the ground as a hammer draws out a nail by the head" [RHAA3/9/392]

Gardeners weeding dug.
[RHAA3/391/137]

Weeding Hooks
An Essex Yeoman owned weeding hooks kept "in the chamber over the servans chamber" [EP19]. Markham mentions hooks and nippers etc to cut or pull up by the rootes thistles [GMFH14]

Scrapple of Iron.
[GMFH59]

Weeding Tonges
3 pairs of weeding tongs worth 2d each were listed in the possesions at Tatton hall [TPP] Markhams wooden nippers to cut or pull up by the rootes thistles were probably the same [GMFH14]

Long small wooden nippers for weeding corn
[GMFH25]

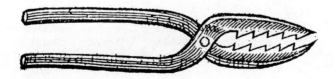

Cutting

Scyth

Scyths were used to cut grass or corn [RHAA3/8/332]. They were found among the possesions of labourers [BP236,253,LP29x2] Husbandmen [BP262], Yeomen [DP180] and Gentlemen [DP59]. An Essex Yeoman owned 2 Sithes, snaths & cradles worth 1/6d kept in the yard [EP19]. Scyths were located in the Soller over the parlour in the inventory of Thomas Turnor husbandman of Cakebole, Worcestershire in 1640. The mowing scyth owned by a Roughmason in Iron Acton near Bristol was almost certainly of this type [IAP5]

The Parts of a Sythe were:

Scyth blade: "that as cuts the grass"
Swath: the long crooked Staff or pole it is fixed unto
Syth hoop and Clat: those that fasten the Sythe to the Swath.
Noggs: the Handles of the Scythe.
Ripp: is that as the mower whetteth his Scythe withal, of some called the Strickle [see under sharpening Strickle]

Sythe shafts bought at local fairs in Yorkshire usually cost 26-28d sometimes 22d or 30d [HB32]. The scyth could be fitted with a rifle,a bent stick on the butt of a scyth handlefor laying the corn in rows, and a cradle,a wooden three forked instrument on which the corn is caught as it falls from the scyth [TT33,324,331]

Grass Scyth
[RHAA3/332/13].

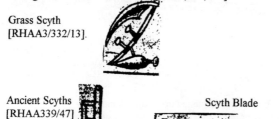

Ancient Scyths
[RHAA339/47]

Scyth Blade

Briar Scyth

One was owned by a Miller near Bristol [IAP38] Tusser refers to a brush scyth [TT33]

Sickle

Sickles had the edge toothed like a saw [RHAA3/7/327] and were "used to reap or cut down corn used by reapers or shearers [RHAA3/8/333]. They were located in the Soller over the parlour in the inventory of Thomas Turnor husbandman of Cakebole, Worcestershire in 1640.

Best recomended that "in choosinge of a sicle, yow are to hold them against the light, and are to see that they bee well toothed; and if soe bee they wante noe teeth, yow neede not care howe small the teeth bee: yow are likewise to minde that the be large and well casten." [HB43] Sharpening or grinding was to be done on the smooth not the toothed side of the blade [HB43]

Sicles
[RHAA3/333/17]

Sicles cutting corn

Similar tool used to prune tree suckers.
[GMEH]

Reeping hook
These are occasionally mentioned [DP65,53]. 4 were owned by a Roughmason in Iron Acton near Bristol [IAP5] and 3 by a husbandman [IAP7]. They are probably normally lumped together with other tools and may be the same as sickles. A blacksmiths stock included 43 reaping hooks valued at 9/- or 2.5 d each [BP239]

Hedging Hook or Pruning Hook
[RHAA3/332/14]

Staff Hook
1 was owned by a Husbandman [BP262] this may be a long handled hedgeing hook.

Crome
Stick or handle with a hook at the end used in clearing ditches [TT324,33]

Hatchet

These were quite common and found in the posesion of Labourers [LP29], Husbandmen [BP235,256x2,262, LP26,IAPx3], Yeomen [DP54,180] and a tailor [IAP4x2] amongst others. A Norfolk blacksmith had a hatchet valued at 10d [WP28] and a Gloucestershire husbanman one worth 8d[IAP10].

In one Devon farm the hatchet was found in the milk house [DP140]. One was located in the hall of Thomas Turnor husbandman of Cakebole, Worcestershire in 1640 another kept in the close by a Gloucestershire Husbandman [CP1]

Hatchet
[RHAA3/380/5]

Hatchet Head
[RHAA3/380/6]

Axe

These were found in the posesion of Gentlemen [DP59], Yeomen [DP54x2], Husbandmen [BP235,256,262,IAP16,19x3] and labourers [BP236]. One was found in the back room of an Essex Yeoman's house in 1638 [EP14]. Two were located in the chamber of Thomas Turnor husbandman of Cakebole, Worcestershire in 1640.

Husbandman Felling Ash trees withan axe
[KPNM]

Carpenters Axe
[RHAA3/380/7]

Axe in use

Moringe Axe
1 moringe axe was owned by a Yeoman [IAP34]

Bills
These were quite common. One was found in the back room of an Essex Yeoman's house in 1638 [EP14], two were located in the chamber in the inventory of Thomas Turnor husbandman of Cakebole, Worcestershire in 1640, another was kept in the close by a Gloucestershire Husbandman who also had a billhook[CP1] while a labourer from the same county owned 2 [CP23] and a Leicestershire husbandman one [LP26]. The Devon trained band pioneers were supposed to have had for every 10 pioneers12 bills to cut wood [Cot16].

Hedgeing Bill
A Gloucestershire husbandman kept a hedgeing bill along with a billhook and 2 hatchets in the kitchen [CP8] They seem quite common in the area [IAP6,7,10,19x2,51]. A husbandman owned one worth 8d [IAP10]

Black Bill
One black bill was included next to a yeoman's armour and gun and may have been considered a weapon [DP34] but another was listed with just after husbandry tools [DP86].

Watch Bill
Owned by a shoemaker [BP313] proably for night policing duties.

Billhook
One was kept in the close by a Gloucestershire Husbandman who also had a bill [CP1]. Another kept a billhook along with a hedgeing bill and 2 hatchets in the kitchen [CP8] others also kept both bills and billhooks [CP5] and these were clearly different items with differing uses. The value of one billhook along with an axe was 1/- [DP194]

Handbill
One was owned by a Leicestershire Labourer [LP29]. This may be the same as a billhook.

Cleaver

A Norfolk blacksmith had a cleaver valued at 8d [WP28] another was owned by a Gloucestershire Yeoman [IAP34][DP59]. Holmes tlist under husbandry "a Woodmans Cleaving Knife, By the strength of it and weighty blows to force it in, great trees are rent, and cloven through, the length of the hadle assisting to prise and forcing them to fly open" it resembled a pioneers pick but differed by "haveing an edge on the inner side, with a socket for the hame or staile to be fastened in; and being much more streighter within. The roundness of it only appearing on the outside to make the back stronger and the point or end the sharper" [RHAA3/392] This is probably a rather enthusiastic description of the "Iron Frower [Froe] for cleaving lath" mentioned in Tussers list of husbandry tools.

Cleaver
[RHAA3/391/135]

Saws

In one Devon farm 2 saws were found in the milk house [DP140] a Gloucestershire husbandman kept his in the loft over the malthouse [IAP51].

Hand Saw

Common [DP59,193,IAP16,19,7etc], one was kept in the close by a Gloucestershire Husbandman [CP1] 4 hand saws were amongst the items to be carried by pioneers in Devon [Cot16]. Possibly the same as the short saw Tusser recomended to cut logs with [TT32]. The hand saw was used "to cut or slit small Timber, as Boards, Spars, Rails, &c, though indeed tha Saw cannot properly be said to cut or slit, but rent, break, or tear away such part of the Wood as the points of the Teeth strike into." [RHAA3/363]

Hand Saw
[RHAA3/380/1]

Tennant Saw

These occur in a number of Gloucestershire inventories for Shearmen [IAP13] and husbandmen [IAP16] in the case of 2 husbandman tenant saws and hand saws are listed showing these are different items [IAP19,16] 2 tent saws, probably tennant saws, were amongst the items to be carried by pioneers in Devon [Cot16].

Tennant saws had relatively small blades and were supported in iron or wooden frames as illustrated. They were optionaly termed bow saws or frame saws [RHAA3/365]

Tennant Saws
[RHAA3/380/2,3]

Whip saw

"Whip saw of some termed a framing saw is a long saw used between two persons to saw great pieces of timber or other stuff" which was too large for other saws. The timber was laid on a tressel and the men alternately thrust and pull. [RHAA3/368]. 1 whip saw was kept in the loft over the kitchen and valued at 12d [CP8]. An Essex Yeoman owned a two hand saw kept "in the chamber over the servans chamber" [EP19]. This is posibly the same type of tool as the long saws Tusser recomended for cutting logs [TT32].

Whip Saw
[RHAA3/380/28]

Thart Saw

A crosscut saw. A Yeoman owned both a thwarte saw and a hand saw indicating they were different [IAP34,DP54] 1 thirte sawe cost 6d [CP24]

Arme Saw

Found in the "Sope Chamber of one Devon farm [DP143]

Sheers
One pair of sheers valued a 3d
One set was owned by a Yeoman [IAP34] another by a miller [IAP38]

Sheep Sheers
A husbandman had two pairs of sheep sheers valued at 1/4d or 8d each. [CP24]. A
blacksmiths stock included 7 pairs of sheep sheers valued at 6 d each [BP239]

Sheep shearers usually brought two pairs of shears with them when employed in North
Yorkshire and sharpened them with a whetstone [HB21]

Shepherd sheared by his sheep 1642

Heave Chopper
Owned by a Husbanman [IAP7]

Spliting

Hay Hook
used "for the pulling out of Hay made either in a Rick, Stack or Mow, when they are about to fodder their Cattle and Beasts [RHAA3/8/334]. A Leicestershire husbandman kept his hay hook in the barn [LP27]

Hay Hooks
[RHAA3/334/21,22]

Cutting Spade
Sharp cutting spade to cut the hay mow [stack] for deviding hay [TT33]

Wedges and Beetles
Iron wedges appear in various numbers, 2 [DP107,59,BP235,256,CP1,IAP22], 3 [DP180,104], 4 [DP196], 5 [CP24,IAP19], 6 [DP164,54], 7 [DP55] and often accompanied by a beetle [DP180,164,193], probably a large handle fitted into a section of tree branch to make a large wooden mallet the size of a sledge hammer. In some cases these were specified as ring beetles meaning one bound round with rings of iron [DP164,107,102] almost certainly not on the striking face but on the other circular sides, to retard splitting. Using an iron headed sledge hammer on modern iron wedges causes them to bend and feather or split off on the upper edges. Period iron was probably less resistent and probably broke up even faster The head of the wooden beetle being softer than the wedges would break up first and was probably relitively cheaper to replace. This may also be reffered to as a Dovercourt Beetle [TT32]There is a refference to a sawe bittle and wedges [DP193]

They were found among the posesions of labourers and husbandmen as well as better off farmers [BP235,236]. In one case wedges were stored in an upstairs chamber [DP196]. 5 were valued at 2/8d total [CP24].

Sharpening

Strickles

The best strickles are those that are made of froughy, unsesoned oake; yow may [one] for 1d, but a good one will cost and is worth 2d...hammer to pitte the strickle with to make it keepe sande [HB32].The worker also kneeded a sand bag and sand which was bought by the penyworth [HB32].

Rubstone and Sand
Probably the same as a strickle.

Whetstone

Among the possesions of a Devonshire husbandman 1 whetstone which with an Iron rake was valued at 13d [DP70]. Sheep shearers sharpened their shears with a whetstone which was small enough to fit in a pocket.[HB21].

Grindstone

These were much more complex pieces of equipment. Grindstones occur in various probates. One was in the buttery [DP140] and another in a Yeoman's larder [DP55] while in 2 case they were valued at 2/6d [DP130,111]. One was given as "a grindinge stone with a Breache" [DP103] probably in a courtyard while another "grinding stone with a trough in the Courte" seems certainly to have been outdoors [DP54] A Gloucestershire husbandman kept his at the back door. A Norfolk blacksmith had a "grindstone cranck and trowe" valued at 3/6d [WP28]

The stone was revolved with a handle in a bath probably of water.

Evelyn

THE SCOTS HOLDING THEIR YOUNG KINGES NOSE TO Y GRINSTO

Jockie

Wheelbarrow
Dung was carried in "little drumblars or wheel-barrowes, made for the purpose, such as being in common use in every husbandmans yard" [GMHM204] and they are certainly not uncommon in probate inventories [DP86,87 etc]. One Devon farmer had 2 wheelbarrows in his barn valued at 5/- [DP195].

[RHAA3/7/328,8/344/85]

Evelyn

Handbarrow
Bearing Barrow, Barrow, Hand barrow.

Burdens are carried on it between two persons, supported by hands as opposed to a wheel barrow.[RHAA3/8/344]. One farmer had 2 hand barrows as well as a wheelbarrow [DP110] while another 2 wheelbarrows and a handbarrow [DP72]. "a whilbarrowe & a peare of hannabordus & hands" [DP23] indicates that the hands may have been detachable poles along the sides.

[RHAA3/344/86]

Well Equipment
Wells involve the vertical transport of water. Entries occur for well buckets, rope and chain. [DP180,111] or just one bucket with an iron chain [DP42] plus on one farm there were references to a well turner and a well chain [DP196]

Buckets
While most buckets were probably made of wood one Yeoman had 3 leather buckets 4/- entered just before 2 new well ropes 4/- [BP248].

Milk Pail
A milk pail was listed just after 2 buckets in one Devon inventory implying a clear distinction between them. [DP164]

Pulleys and Ropes
There are a few references to pulleys and ropes. This may be due to a very low value on the items. [DP180]

Gathering, Spreading and Flattening

Rakes
A Leicestershire husbandman kept his rake in the barn [LP27].

Hay Rakes
Hay rakes were made entirely of wood, usually sallow or ash. "Hay-rakes may be bought at Malton for 22d a dozen; they usually have 15 teeth a peece, and are all of saugh, bothe shafte, heade, and teeth; sometimes the heades and teeth are of ashe." [HB34].

Mead Rakes An Essex Yeoman owned 4 meade Rakes kept "in the chamber over the servans chamber" [EP19]. These were probably meadow or hay rakes.

Garden Rake
Rakes are listed among the few tools possesed by a widow and a maiden [IAP18,23] in the maidens case it was specified as a garden rake and was kept in the entry. Rakes were probably of differing teeth sizes as Markham recomended using a fine rake on seed beds just after sowing [GMHM2/6]

Garden raked seem to have iron heads. 1 Ireracke [Iron rake] for the gardenes [gardens] belonged to a Gloucestershire labourer [CP23] while Markham reffered to "your Garden Rake of Iron" [GMHM206,4]. 1 Garden rake listed in Devon was valued with some ropes at 20d [DP54] A blacksmiths stock included 5 rake heads valued at 1/4 ie 16d or 3.2d each [BP239]

[THGL]

Dew Rake
An Essex Yeoman owned a dew rake kept
"in the chamber over the servans chamber" [EP19].

Sweath Rake
Used to gather lose ears of corn in the field.
This job was reserved for the most unfit [HB43]

Barley Rake This had iron and/or steel teeth

Mauls

Garden Mauls
"which are broad boards of more than two foote square set at the ends of strong staves, the earth shall bee beaten so hard and firme together, that it may beare the burdern of a man without shrinking" used for the preparation of beds in ornamental gardens [GMHM206].

Flat Mauls or Bettles
Lighter broader and flatter than clotting bettle a thick ash board more than 1 foot square and 2 inches thick [GMFH13]

[GMEH13]   [RHAA3/391/136]

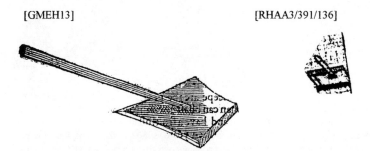

Mauls or Clotting Bettles
Mauls or Clotting bettle. Good strong clotting bettle, breaks lumps of earth to Dust [GMFH12,HB138]

[GMEH12]

Forks

One may buy...at Malton shorte forke-shaftes, made of seasoned ashe, and quarter cliffe for 2s or 22d a dozen [HB34]. A Leicestershire labourer owned 2 forks [LP29].

Pike
Farms oftened owned a number of pikes. 3 were kept in the loft over the malthouse by a husbandman [IAP51] and 4 in a barn by a Yeoman [IAP39] while a labourer owned 3 [IAP36] and a tailor 2 [IAP4]. A staff for a pike cost 6d [IAP67] as did a pike [IAP10]. This was probably an alternative name for fork at in Pikefork.

Forks came in a wide variety of types:

Pitchfork, or Pickel, or Pickfork
"much used in husbandry for their loading and stacking of hay and corn... Some are made with a socket for the staff to go into, but the general way of making them is with a tang and a shoulder, to go into a hole made in the staff hooped about with Iron" [RHAA3/8/331]. The metal head of the fork was known as the graines [RHAA3/7/328]

Pitchforkes were kept in stables [TT31] A Leicestershire husbandman kept 3 pitchforks in the barn [LP27]. An Oxfordshire shoemaker had 5 picforks [BP313,EP8]

Socket Pitchfork or Socket Pickel
[RHAA3/9/390]

Short Pitchfork
In Barn [TT31]

Long Pitchfork
In Barn [TT31]

Light Pitchfork
Light Pitchfork and tough with Cart [TT32]

Straw Fork
In Barn [TT31]

Hay Fork
[TT33]

Corn fork or Pikes
This differed from the hayfork. One devon Yeoman had 4 hay pikes and 4 corn pikes in the chamber over the hall [DP55]. 2 corn pikes were valued at 1/2d. [IAP9] Two were kept in a Devon stable .[DP]

Dungfork

Alternative names Worthing fork, Yelve or Evill [RHAA3/8/337]. One Devon farm listed 3 evells [DP65] another 2 dung evells [DP180]. Thomas Tusser expected there to be one in the stable [TT31] and one was recorded so in Devon [DP44] where a yeoman had 5 in the stable or the hay loft above [DP55]. One was located in the hall of Thomas Turnor husbandman of Cakebole, Worcestershire in 1640 while a Leicestershire husbandman kept his muckforke in the kitchen [LP27]

Type A

Used to Yeron or Meron ie clean out stables and Cowsheds of beast dung casting it onto the dunghill, and to load it onto a cart for movement to the fields.[RHAA3/8/337]

The Parts

Bar or Cross Bar
Tangs or forks
Socket for the stall to go in
The Stall
Kaspe: the top part the man holds. [RHAA3/8/337]

Type B

There is another sort of these dung Forks with only two ends or points, and without a Kaspe or potent on the head ...a dung iron or a yelve iron with two ends [RHAA3/8/337]

Type C

Square Worthing Fork, Square Dung Fork, Socketed Pickel Double Barred, Foot Yleve. With two points and a square metal base it enabled a man to put his foot on it to force it into the dung hill [RHAA3/9/392]

Dung Fork with points.   Dungfork
RHAA3/391/136]   [RHAA3/337/43]

Dung Fork Heads
[RHAA3/337/44]

Beating

Flail or Theshall

The Prostant Flayl

This flayl it was made of the finest wood
Well lined with lead, and notable good
Two handfulls of Death with a Thong hung fast
With a moving head both stiff and stout,
Found by the Protestant Joyner out, [DSBM]

Parts of the Flail:

Hand Staff: that the thresher holds it by
Swiple: The part that strikes the corn
Caplings: the strong double leathers made fast to the top of the hand staff, and the top of the swiple
Middle Band: the leather thong or fish skin that tyeth them together [RHAA3/8/333]

A flail consisted of two truncheon like pieces of wood joined by a flexible link. At the end of each piece of wood was a double thickness piece of leather fastened on in a loop. These were conected with a leather thong or piece of fishskin. Fishskin can be tanned to form leather strong enough for boots and this may have been done to strengthen the skin. The "truncheon" used to hit the grain might be weighted by filling the centre with lead.

Brake

A Gloucestershire labourer owned "a bracke to poule Bromme [IAP36] This was probably a device through which broom was pulled to strip it possibly prior to making brushes. A husbandman in the area also kept one in the loft over the malthouse [IAP51,10].

Sorting

Sieves

One husbandan had 9 sieves in addition to a malt sieve and two searches [these were in proximity to a peck and two half peck measures and a bag]. [IAP19] An Oxfordshire labourer owned a sieve [BP229]

Coal or Lyme Sieve

Wide square holes in the bottom which a mans finger can be thrust through each hole. The bottom was made of split wood. [RHAA3/8/337]

Garden Sieve

The bottom was made of strong wire squares as large as the coal sieve. [RHAA3/8/337] Soil could also be sieved through a wattle hurdle or round or square sives with bases of plated sticks or wire. [PP3]

Sieve or riddle
[RHAA3/337]

Riddles

Large bottomed sieves such as the coal and garden sieves could also be termed riddles. [RHAA3/8/337]. A ridling sieve was used to scatter earth over freshly sown garden seeds [GMHM6]. A riddle was listed among the possesions of a Husbanman [IAP7]

Reeving Sieve

The Reeving Sieve is to clense corn at the time of Winnowing from the dregs of chaff, and the small Seeds of Tarres and Lintels which are in it, which is termed Reeving of Corn [RHAA3/8/331]. A ryen seave was listed among the possesions of a Gloucestershire Husbanman [IAP7]

Malt Sieve

A malt sieve was listed among the possesions of a Gloucestershire Husbanman [IAP7].

Meal Sieve or Bolting Sieve

The bottom was made of a kind of fine hair cloth course enough to allow the meal or fine flour to pass through but retain the bran. [RHAA3/8/337]

Fine Sieve or Silk Sieve

A fine tiffany bottom allowing only the pure dant of ground corn to pass through retaining all tpes of bran within the rim. [RHAA3/8/337]

Searce or Searcer

A fine sieve with a leather cover on the top and bottom of the sieve rim to keep the dant or flour of any pulverised substance so that none was lost in the searcing. [RHAA3/8/337]

Winnowing sheet, and cloth

The most common combination of sorting equipment in farm inventories was the winnowing sheet usually accompanied by a sieve or haire cloth and baggs often termed corn bags. [DP195,120,117 etc]. One Yeomans inventory casts a clearer light on their use "In the chamber bags for corn the winding sheet and sieves to winnow the corn 20/-, the ranges and clensing sieves 3/- a peck and a half peck 12d" [DP54]. The winding or winnowing sheet was for tossing up the threshed grain to allow the wind to separate the heavier wheat from the chaff. Threshing was probably usually undertaken through the winter when there was spare labour [HB] and the usuall location seems to be a barn with opposing doors which could be opened to allow a through draft in a constant direction without danger from showers. Once the main separation had taken place the sieves removed weed seed and grit etc. The grain was probably then measured by pecks or half pecks by volume [2 gallons, 4 pecks make a bushell] and packed in sacks for transfer to long term storage or possibly to the mill.

Winnowing sheet
A winnow was listed among the possesions of a Gloucestershire Husbanman in the backhouse [IAP7] and a Yeoman kept his in the barn [IAP39] in Devon they appear to be termed wyndinge sheetes [DP32]

Hair Cloth
1 Hair Cloth 6/- [DP67,DP72] Chamber over brewhouse, 2 harecloth [DP55]

Canvas Sheet
Winnowing sheet and canvas sheet [DP65]

Packsheets
Found with winnowing kit in Devon [DP138,112,164]

Dewe sheet
Dewe sheet [DP103] Yeoman, Yarn Sack and Dowe Sheet [Husbandman [DP104]

Plitch clothes
Plitch clothes [DP42] 3 plitch bags [DP59]

Market Cloth
Market Cloth belonging to a Yeoman [DP103]

Measures

Hoop, Strike or Bushel

A measure is also called a Hoop, Strike or Bushel. Middle, Bottom or double measures were distinguished from a single. "Has a hoop in the middle to show it can measure a double or full measure or a single or half measure". [RHAA3/8/331] A hoop, strike or measure consisted of 9 gallons or 4 pecks. In some places 8 gallons was called a bushel. 4 measures of corn or 5 of oats made a bushell in Cheshire [RHAA3/337]

A number of Devon farms had a bushel and a half bushel measure. [DP112,110]. One Devon farm had a bushel bag [DP] while 2 great bags of rye being 12 strikes were listed in the inventory of Thomas Turnor husbandman of Cakebole, Worcestershire 1640. A strike was listed among the possesions of a Gloucestershire Husbandman in the chamber over the kitchen [CP1] and one owned by an Oxfordshire labourer [BP229].

Measure [RHAA3/337]

Strickle

"a staight bord, with a staff fixed in the side, to draw over the corn in measuring, that in exceeds not the height of the measure" [RHAA3/8/337,339]. This was drawn accros the top of the filled measure to leavel the contents off.

Scales

They consisted of a ballenced beam and scales. The beam seems to be normally of iron when they are listed in inventories. 2 beams, scales and weights appear in one Devon probate [DP55], 1 iron beam and skales worth 5/- in another [BP248] and 1 little pair of scales with an iron beam belonging to a husbandman in Gloucestershire in another. [IAP19]

Weights

One large farm in Yorkshire weighed its wool "in single stones, because the scale would holde noe more but a stone" [HB31] The weights available on the farm were described as:

a two stone weight with a ringe, beinge of leade, rounde and sealed
a rounde halfe-stone or 7lb. weight ringed
two flat halfe-stone weights, sealed and marked with the flower de lyce and crowne
a fower pownde weight, flatte, and marked with EL and a crowne havinge a figure of 4
a two pownde and a single pownde, three square and sealed
two rownde halfe powndes

Of these the two flat 7 pound [half stone] weights were used to weigh wool and some smaller weights [HB31].

Small weights were listed among the possesions of a husbandman who also owned 2 pairs of brass scales. These may be for medicinal quantities rather than farm devices [IAP19].

Bags

According to Holmes sacks could hold 4, 5 or 6 measures of corn, Poughs 1,2 or 3 measures and bags 1,2 or 3 pecks [.25,.5.or .75 measures].[RHAA3/336] This is partially confirmed by a refference in Tusser to ten sacks each holding one croome or 4 measures.

However probates indicate the terms bags and sacks may be interchangable. 2 great bags of rye being 12 strikes listed in the inventory of Thomas Turnor husbandman of Cakebole, Worcestershire in 1640 would contain 6 measures and therefore large sacks by Holmes reconing. One reference talks of 3 bags or sacks worth 12d. Bags varied in price, the former costing 4d each, a Yeoman's sack 1/- [IAP20], and in another case 6 bags 10/- or 20d each [IAP17]. Bags apparently varied in size and use. One refference is to malt sacks and bushel sacks [a bushel is 4 measures in some areas]. In one case they were stored in the chamber over brewhouse [DP55] although barns seem most common [DP72].

They are usually found with sieves, peck measures winding sheets and often hair cloths in Devon.

Sack of Corn
[RHAA3/337/37]

Climbing

Ladders

These are widespread in inventories often kept in the barn [IAP52,62x2]. A Leicestershire husbandman kept his "lather" in the barn [LP27] and a Devon man one in the Lynhey and another in the barn [DP44]. A Devonshire woman had 2 in the stables [DP140]. A short ladder and two long ladders were kept in the chamber over the kitchen by a Gloucestershire Husbandman [CP1] and another had 1 long and 1 short [IAP7]. One widow had 3 ladders and a "laddersheed" [DP110] possibly a shed for keeping the ladders in. Tusser talks of

"Long ladder to hang al along by the wal
to reach for a neede to the top of thy hal"

Indicating another storage method.

One ladder cost 10d while 2 ladder peeeces on the same farm valued at 2/- [IAP10] another valued at 6d [DP196] while a Devon gentleman had ladders worth 6/8d [DP198].

Taper Ladder

Taper ladder of staves. a kind of ladder or cheese rack having one end wider or broader than the other...a ladder broader at the foot than at the top part. [RHAA3/9/390]

Cart Ladder

Owned by a carter [IAP43]

Carpentry Tools

Carpentry tools appear not infrequently in farmers inventories and it is probable that many farmers were equiped to make or repair wooden tools, buildings or equipment. A detailed list survives of one husbandman's carpentry tools:

1 draw knife	1 spoke shave	1 tennant saw
2 hammers	2 pairs of pincers	1 hand saw
2 chisels	2 borers	

Hammer
These are mentioned in a number of probate inventories [DP65,59] Tusser mentions a lath hammer presumably for nailing on laths for plaster [TT33] There is also refference to a stone hammer probably for dressing stone [IAP67]

Pair of Pincers
Also found in a Devon probate [DP65].

Clavestock
This appears in Tusser list of farm equipment as a tool desired by carpenters. It is identified as a choper for splitting wood and may be an alternative term for a froe or cleaver although Tusser also mentions a froe. [TT322]

Rabetstock
This appears in Tusser list of farm equipment as a tool desired by carpenters. It is identified as a rabbet plane.[TT331].

Adze
One was listed among the possesions of a Gloucestershire husbandman [IAP16]. Tusser lists an adze as being usefull for making troughs for hogs, presumably hollowing out tree trunk sections.

Chizel
Also appears in a Devon probate [DP59]

Boorer and Augers
These were probably alternative names for the same item. One was kept in the close by a Gloucestershire Husbandman [CP1] They appear not infrequently in Gloucestershire husbandmans possesions [IAP16], one owning 5 [IAP7] and another 2 [IAP19].The Devon trained band pioneers should have had for every 10 pioneers 10 berriers (augers) of several sizes [Cot16].

Augers had a wooden handle and shank and a bit. They were used with both hands either beneath the worker or at chest height to make large round holes. Bit diameters varied from .25 inch to 4,5 or 6 inch [RHAA3/365].

Miscelaneous

A number of miscelaneous items are mentioned:

Marking Irons
[IAP34,38x2] Possibly branding irons for livestock.

Brush
1 wanting brush 6d,1 Brush 6d [IAP20] Presumably for sweeping the barns or yards.
Tusser also includes a wing among threshing equipment and goose wings seem to have
been used as grain or chaff brushes. He also lists a broom in the same area and another
along with stable equipment [TT31].

Cutting Knife
1 old cutting knife was among a millers possesions [IAP38]

Sprangs
Spring traps possibly for moles, humans or other predators [DP196].

Masonary Tools

Trowel
1 trewell was listed among a husbandmans possesions [IAP19] possibly a builders trowel
for masonary work.

Hod or Tray
Probably for carrying stone or brick. [TT33]

Tray
Wooden tray: used for mixing garden seeds with earth before sowing [GMHM6]

Baskets
The Devon trained band pioneers should have had for every 10 pioneers:12 small baskets
to carry earth [Cot16].

Fraile or Skep
1 fraile and seves [Husbandman [DP104]. Tusser mentions a skep both seem to be baskets
made of rushes or straw [Web,TT332] These seem related to threshing activities.

Mole Spear
Sharp Mole Spear with Barbs.[TT]

Sheep Crooks

Vital to shepheards, these are clearly shown in woodcuts. The heads appear to be made from sheep's horns:

Hop Pole Remover

This final device was for removing old hop poles without unduly disturbing the soil. It use a pair f iron pincers at least 5 feet long and a "clasping hook"to hold the teeth together once they were fastened to the pole as close to the ground as possible. A piece of wood was then placed as a fulcrum and the pole leavered out of the ground [GMFH]

Thomas Tusser's list of the tools and equipment on a small farm wereproduced in 1580
The list in Randal Holmes Academy of Armory produced pre 1688 was clearly based on
Tusser but slightly modified.

Randal Holm:

Things necessary for a good Farm or Dairy.

....

In the Barn
Barn well locked
Pitchforks long and short
Straw fork, Rake and Ladder
Broom, Wing, Winnow Sheet,and Sack with a Band.
Shovell, Peck, Bushell and Stricles.
Reeving Sieves
Seed Corn, Seed Hoppet.

In the Stable
Stable well planked, locked and Chained.
Strong Walls and well Lined.
Good stall, Cratch or Rack, good hay and Litter
Manger Chaff and Provender
For Dungfork and Hayhook
Sieve Skep, Bin Broom and Pail.
Handbarrow, Wheel-barrow, Shovell and Spade
Spunge, Curry Combe, Mane Combe, Whip.
Hammer, Nails, Buttrice and Pincers.
Bridle, Saddle, Pannel, Pad, Pack Saddle
Waunty, Whit-leather and Nall.
Slips Collars, Harness, Halter, Headstall and Cord.
Crotches, Pins, Apron and Cisars.

In the Cow House

...

A Fork or Evill or Yelve.

In the Cart Shed

...

Piercer, Pod, Pitchfork or Pikell.
White or shave, whiplash, Goad and Rope.
Pulling Hook, Hand Hook, Sickle and Scythe
Tumbrell, Dung Crone, Pick Axe.
Mattock, Bottle and Bag.
Plough, Chain, Coulters, Shares and Sucks

Ground Clouts, side Clouts.
Plough Bettle, staff and Slade.
Oxbo?es, Oxyokes, Horse Collars.
Oxe-Teeme and Horse-Teem.
Rake Iron Toothed, Harrow, Weeding Hook.
Hay hook, Sickle, Fork and Rake.
Bush Scythe, Grass Sythe, Rifle and Cradle.
Rubbing Stone, Sand Whetstone and Grindlestone.
Skuttle or Skreine or Sieve.
Tar, Tar-pot, Sheep Mark, Tar Kettle.
Shearing Shears for Sheep.
Yoke for Swine, Twitchers or Rings.
Long and Short Ladders and a Lath Hammer.

In the Farmers House.
Trowell, Hod and Tray,Scales, Beam and all sorts of Weights.
Sharp Mole Spear with Barbs.
Sharp cutting Spade to cut the hay Mow.
Soles Fetters, Shackles, Horselocks, Padlocks.
Claverstock, Rabbet stock, a Jack to Saw upon, and Pinwood timber.
With a Didall and Crome to drain ditches.
Hatchet, Bill, Axe, Ads and nails of all sorts.
Iron Frower, and wedges to cleave laths and Wood.
Saws long and short, Beetle and levers with a roll for a Saw-Pit. [RHAA3/244]

Tussers list includes the following tools:
Barn Furniture
Gofe ladder
Short and Long Pitchforks
Flaile, strawfork and rake, with a fan
Wing, Cartknave and bushel, peck strike
Casting Shovel, broom sack with a band.

Stable Furniture
Pitchfork, Dungfork, sieve, skep, bin
Broom, pail for water
Handbarrow, wheelbarrow, shovel and spade
Currey Comb. maine combe and whip.
Buttrice, Pincers, hammer and nails.
Aperne and siscors for head and tail.

Cart Furniture
Cart lader and wimble with percer and prod
Wheel ladder for harvestlight pitchfork.
shave, whiplash well knotted

Ten sacks each holding one croome. [crome =half a quarter]
Pulling hook for bushes and broome
Shovel, Pickaxe, Mattock with bottle and bag

Husbandry Tools
Grindstone, Whetstone, hatchet and bill
Hammer, English Nail
Iron Frower for cleaving lath
Roule for a sawpit

Short saw, long saw to cut logs.
Axe and Adze
Dovercourt Beetle and wedges with steel.
Strong lever to raise block from wheels

Plough beetle, ploughstaff

Barley rake toothed with iron and steel
brush scyth and grass scyth with riffle to stand.
A cradle for barlie runstone and sand
Sharp sickle and weeding hook hay fork and rake.
meake for pease and to swing up brake.

Sort rakes for to gather up barley to bind and greater to rake up such leavings behind
A rake for to hale up fitchis that lie a pike for to pike them up handsome to dri

Skuttle or skreine to rid soil from corn
Sheep shears
Fork and Hook to be tampring in claie
Lath hammer trowel hod or a traie.

Long ladder to hang al along by the wal
to reach for a neede to the top of thy hal
Beame scales with the weights that be sealed and true
Sharp mole spear with barbs

Sharp cutting spade for dividing mow.
Skuppat and Skavel for marsh men
Sickle to cut with a didall and crome
for draining ditches

Claverstock and Rabetstock carpenters crave.
Seasoned timber for pinwood
Jack for sawing firewood on. [TT31-33]

ABBREVIATIONS

This work uses the Stuart Press standard abbreviations system . The Letters identify the work, numbers refer to the page number unless otherwise specified. If the number is subdivided for example 2/3/164 This means page 164 of the third volume of the second book, or whatever equivalent subdivisions the work is organised in. In some period works the text is numbered in mini chapters rather than pages and a plain number will refer to this.

AHT	Annals of Henry Townsend, Worcestershire Record Soc???
CJ	Commons Journal
CSPD	Calender of State Papers Domestic, number reffers to date of entry.
CSPV	Callender of State Papers Venitian.
CSPI	Callender of State Papers Ireland.
DP	Devonshire Probate Wills 1531-1660 [the number is the probate number not the page number]
DSMB	David Stell. Ballads and Music of the early 17th Century. Stuart Press 1994.
E	Thomason Tracts, University of An Arbour Microfilms.
EA	John Evelyn: Acetaria 1699
EF	Hillary Spurling: Elinor Fettiplace's Reciept Book 1604-1647? Penguin 1987
EK	John Evelyn: Kalendaria 1664
EP	Farm and Cottage Inventories of Mid-Essex 1635-1749
ESHA	Early Stuart Household Accounts: Edited by Lionel M. Munby, published 1986 by Hertfordshire Record Publications.
Ex	Exchequer Papers, Public Record Office
FCEP	Farm and Cottage Inventories of Mid Essex.
GH	Gerrards Herbal 1633
GMCG	Gervase Markham: Cheap and Good Husbandry 1631.
GMEH	Gervase Markham: The English Huswife 1648
GMFH	Gervase Markham: Farewell to Husbandry 1648
GMHM	Gervase Markham: The English Husbandman 1635
HB	The Farm Books of Henry Best 1640. Camden Society 1848?
HLA	John Burnett: A History of the Cost of Living. 1969. Pelican.
IAP	The Goods and Chattles of our Forefathers. Frampton Cotterell and District Probate Inventories 1539-1804. Edited John S. Moore.
LP	Leicestershire Probate Inventories
MDCJ	Michael Dalton: The Country Justice, 5th edition 1635.
OED	Oxford English Dictionary
PP	Parkinson: In paredisi in Solei...
RHAA	Randle Holme: Academy of Armory 1688
SP	State Papers, Public Record Office
THGL	Thomas Hill: The Gardeners Labyrinth 1652 ed First published 1577
TP	Tradescants Paintings, Bodleian Library, Oxford
TPP	Inventorie of John Egerton of Tatton in the County of Chester 4/4/1614
TT	Thomas Tusser: Five hundred points of Good Husbandry 1580. Oxford University Press 1984.
Web	Websters New International Dictionary 1927. G. and C. Merriam London
WL	William Lawson: A new ORCHARD, and GARDEN (1648)
WP	Wymondham Probate Inventories 1590-1641

Publications by Stuart Press

Stuart Press is the publishing division of Historical Management Associates Ltd. All books may be ordered by post from Historical Management Associates Ltd, 117 Farleigh Road, Backwell, Bristol. BS19 3PG. Please enclose an A4 or A5 stamped addressed envelope and payment. Unless otherwise specified all books are A5 size and relate to the period 1580-1660. Send SAE for full catalogue.

English Civil War Battles Series

Battle books usually cover the background to the battle, a detailed account of the battle and details of the units involved. They also include contemporary accounts.

Living History Reference Series

Prices and Wages 1630-1650: R. and J. Hugget and S. Peachey. Common military and civilian wages and prices in war and peace. ISBN.185804006X. 32p. £3.00

Measures and Dates 1580-1660: A researchers quick guide: Stuart Peachey. 12p. £2.00

17th Century Woolen Cloth Specification: S. Peachey. The weights, lengths and widths of various cloth as laid down by statute. ISBN.1858040027. 16p £2.00

Bastardie: Stuart Peachey. The consequences of fathering a bastard for the parents and the child's upbringing with samples of the documentation involved. ISBN.1858040019. 16p £2.00

17th Century Sex: Mc Sween and Jones. A guide to sexual practice. ISBN.1858040000. 16p £2.00

Advanced 17th Century Sex: An illustrated guide based on Italian/Dutch woodcuts. 16p £2.00

Lust 1450-1660: Jane Huggett. ISBN.1858040639. 16p £2.00

Witchcraft in Seventeenth Century England: Dr. John Swain. ISBN.1858040485. 20p. £2.00.

Ballads and Music of the Early 17th Century: David Stell ISBN.1858040108. 60p £5.00

The Complete Receipt Book of Ladie Elynor Fetiplace Volume 1: Never before published in full this is the first of a 3 volume transcription of the whole original text. About 90% of the work is household remedies from a country gentlewoman the remainder mainly cullinary. 68p £5.00

The Mirror of Health: Jane Hugget. Diet and Medical Theory 1450-1660 55p £5.

Medicine 1580-1650: Hugh Petrie.

    Volume 1: The Physician's Books Opened: Background and practices of physicians 28p £3.00

    Volume 2: The Surgeon's Box Opened: Surgeons and Surgery. 24p £3

Common Soldiers Clothing of the Civil Wars 1639-1646: Volume 1: Infantry. Stuart Peachey and Alan Turton. The latest research on clothing supplied to the armies and the civilian clothing worn where clothing was not issued. Includes patterns and fabric and colour details. 56p £5.00

Charcoal Burning in the 17th Century: Dr Malcomb Stratford. A practical guide which covers the period sources and instructions on how to go about charcoal burning today in the traditional manner. ISBN.1858040434. 36p £3.00

English Agriculture 1580-1660

Shepherds and Sheep 1580-1660: Rob Stuart. The life and activities of shepherds and the management of their flocks in England from 1580 to 1660. ISBN 1858040647 40p £4.00

Pigs, Goats and Poultry 1580-1660: Rob Stuart. ISBN1858040663 24p £3.00

Hay Making and Meadow Management 1580-1660: Rob Stuart ISBN1858040671 12p £1.50

Fruit Varieties Register 1580-1660: Stuart Peachey. The register includes all period references to fruit varieties likely to have been found on the normal farms and gardens of the period. This includes descriptions of appearance, picking times, keeping and cooking qualities etc where possible. The register also includes all original black and white illustrations of the varieties.

    Volume 1 Apples to Mulberries ISBN185804071X 52p £5

    Volume 2 Nectarines to Walnuts ISBN1858040728 60p £5

Cottage and Farmhouse Gardens 1580-1660: Stuart Peachey . The layout and techniques. 32p £3

Farmhouse and Cottage Garden Plants: Stuart Peachey. The plants grown in period gardens with contemporary illustrations, descriptions, propogation details etc

    Vol 1: Culinary part 1: 60p £5

    Vol 2: Culinary part 2 and Aromatic: 56p £5

    Vol 3: Medicinal and Decorative: 64p £5

Agricultural Hand Tools 1580-1660: Rob Stuart. The owners, costs and form of hand tools. 40p £4

Early 17th Century Food Series

The Gourmet's Guide 1580-1660: Stuart Peachey. The very best of period cookery. The original recipes with modern interpretation. ISBN1858040531. 60p £5

The Tiplers Guide to the Mid 17th Century: Stuart Peachey. A comprehensive guide as to the drink consumed and made, the law on drinking and a brief guide to drinking vessels. Includes many recipes. Identifies many of the wines used as ingredients in cookery. ISBN.1858040078. 84p £5.00

Cookery Techniques and Equipment 1580-1660: Stuart Peachey. £5 each
> **Volume 1: Fuel and cooking methods and equipment** including ovens, roasting, boiling, grilling etc. ISBN.1858040515
> **Volume 2: Food preparation methods and equipment** including grinding, sorting, drying, cutting, colouring etc. ISBN.1858040523

Feast and Banquet Menus: Stuart Peachey. A collection of sample feast and banquet menus from period cookbooks. ISBN.1858040051. 24p. £2.50.

17th Century Vegetable Uses: T. Mc Sween. A brief guide to the vegetables and their uses. ISBN.1858040098. 12p £1.20

Civil War and Salt Fish: Stuart Peachey. An analysis of the food and diet of the common man in the mid 17th century. £5.00 **Partizan Press**

Early 17th Century Imported Foods: Stuart Peachey. The origins and nature of imported foods in England. ISBN.1858040175. 32p £3.

The Book of Cheese 1580-1660: Stuart Peachey. This work is designed to help identify the types of cheese being produced and consumed in England and Wales. ISBN.1858040272. 16p £2.00

Specific types of dish: These books generally provide a comprehensive list of recipes on a particular type of dish such as Salads or Pies. They contain the majority if not all surviving printed recipes with lists of ingredients and modern interpretations:

The Book of Salads. 1580-1660: ISBN.1858040264. Stuart Peachey. 28p £3.00

The Book of Pies 1580-1660: Stuart Peachey. A comprehensive guide to period pies and tarts including pastry making. Includes modern interpretations for the major variants. 2 Volumes each 64p each £5 **Volume 1: Meat Pies and Pastry.** ISBN1858040582.
> **Volume 2: Fish, Dairy, Egg, Fruit and Vegetable Pies, Tarts and Florentines.**

The Book of Roasts 1580-1660. Stuart Peachey. A comprehensive guide to period roasts. Includes sauces and stuffings. ISBN.1858040701. 64p £5

The Book of Biscuits and Cakes 1580-1660: Stuart Peachey 36p £4

The Book of Breads: Stuart Peachey 24p £2.50

Transcribed Cook Books These books generally contain the text and a glossary but no modern interpretation of the recipes.

The Good Huswifes Handmaid for the Kitchen: Edited Stuart Peachey. A transcription of a general period cookery book with brief glossary. ISBN.1858040035. 72p. £5.00.

John Murrell's Two books of Cookerie and Carving 1638: Edited Stuart Peachey
> **Book 1 [Recipes and Menus]:** ISBN.1858040248. 44p £4.00
> **Book 2 [Recipes and Etiquette]:** ISBN.1858040256. 56p £5.00

Delightfull Daily Exercises for Ladies and Gentlewomen: John Murrell. Edited by Stuart Peachey. This work contains mainly banqueting rather than feast recipes, banquet layouts and interesting clues to cooking techniques. 1621. ISBN.1858040329 48p £4.00.

Miscellaneous Recipes Volume 1: Stuart Peachey. A collection of scattered recipes in period works other than cookery books. 16p £2.

The Ladies Closet Opened: Edited by Stuart Peachey. The cookery recipes only from the original work dealing with a mixture of dairy, feast, conserving and banqueting dishes. 32p £3.

Feast and Banquet Service

Historical Management Associates Ltd provide accurate Historical period feasts and banquets. We have run major feasts for over 7 years at sites such as Wells Cathedral and Farnham Castle, as well as acting as advisors to organisations including Hilton Hotels and BBC.